Sheila Gilheany

Illustrated by Brian Fitzgerald

Ten Speed Press

Berkeley, California

TEN SPEED PRESS
P.O. Box 7123
Berkeley, CA 94707

First published in 1996 in the
United Kingdom by
The Appletree Press Ltd.,
19-21 Alfred Street,
Belfast BT2 8DL

CIP data on file with Ten Speed Press

First Ten Speed Press Printing 1996
Printed in the UK
1 2 3 4 5 - 99 98 97 96

Introduction

The constellations were originally thought of as scenes in the heavens. They then became a way of mapping different regions of the sky. This book looks at twenty of the best known constellations, each of which has its own myths and legends as well as holding its own stellar mysteries. They are arranged according to the season of the year in which they can be seen in the northern hemisphere.

Each night, across the skies, we can see stars being born, maturing, and dying. Every type of star imaginable can be spotted, from dwarfs to supergiants, from supernovae to black holes. We can look for other solar systems and wonder, Is there life beyond Earth? We can gaze out beyond our own Milky Way into other worlds and see the gigantic collections of stars that make up the galaxies. We can even look back in time: as the light that we see from the stars takes time to travel to us, so we are seeing the universe as it was in the past. What, in time, will we discover about its future?

Orion

Orion is the mighty hunter who stalks across the dark skies of autumn and winter, forever in pursuit of the beautiful Seven Sisters clustered in the nearby constellation of Taurus, the bull. The giant figure of Orion so dominates the southern part of the sky that he features in the tales and legends of many of the ancient peoples. The Greeks thought of Orion as a hunter or warrior, brandishing his club and shield against the charge of Taurus. The son of Poseidon and the lover of Artemis, his pride was brought low when he was slain by the Scorpion. By Artemis's request to the Gods, Orion is always to be found in the opposite part of the skies from the constellation Scorpius, which only creeps above the eastern horizon as Orion slips below its western edge.

The Orion stars sparkle brilliantly throughout the long frosty winter nights, attracting star-gazers worldwide. His powerful body is outlined with some of the brightest and most varied stars in the heavens. The outstretched left arm is given by Betelgeuse, a dramatic orange-red supergiant. This star, although much older and cooler, is over five hundred times larger than our own Sun and is now nearing the end of its adult life. In contrast, Rigel forms the lower right leg of the Hunter. This young blue-white supergiant, while smaller than its red companion, is twice as hot as the Sun and is the seventh brightest star in the sky. Bellatrix, the Female Warrior, marks his right shoulder while Saiph is situated on the left knee.

Following Orion are his two dogs, Canis Major and Minor, and it is in the former that Sirius, the brightest star in the sky, lies.

Across Orion lies his belt, marked by the three stars Mintaka, Alnilam and Alnitak, or the Three Kings, while deep in the sword region, hanging from the belt, can be found the Orion Nebula. This mysterious, misty patch is the birthplace of a new generation of stars. In this stellar nursery, thousands of light years from Earth, great swaths of dust and gas are falling together under the force of gravity. As the material is crushed by enormous pressures, astronomers can detect the radiation pouring out from these new stars. Perhaps, as these stars emerge slowly from their dusty mantles, future generations on Earth will see new legends in this fertile region of the sky.

Aries

Aries, the Ram, is one of the most ancient of the constellations. In Greek mythology its coat represented the Golden Fleece, which Jason and the crew of the Argonaut sought. The two brightest stars of the constellation have Arabic names: Hamal, meaning lamb, and Sheraton, meaning two. These stars mark the base of the horns of the Ram.

Aries is the first sign of the zodiac. The latter is the collection of constellations through which the Sun appears to pass throughout the year. Zodiac literally means a "circle of animals," and all but one of the twelve zodiacal groups represents an animate object, Libra being the exception. As the planets of the solar system are all orbiting around the Sun, they are also always to be found within the zodiacal ring. Given that the planets were thought to be the gods striding through the stellar constellations, it is not surprising that the cult of astrology was established. Astrologers ascribed certain characteristics to people depending on their zodiacal sign and which planets were in the sign at the time of birth. These characteristics combined elements of

the constellation such as the stubborness of the ram and the regal nature of the lion.

As the Earth spins on its axis it wobbles a little, a phenomenon known as precession. Although this wobble is very slight, over hundreds of years it can clearly be detected. It is for this reason that the signs of the zodiac in astrology no longer correspond with the constellation of the same name. For example, when the Sun is in the sign of Aries, from the middle of March to mid-April, as designated by astrologers, it is really in the region of the sky known as the constellation of Pisces.

The notion of astrology foretelling the future dates back to the use of the stars as a means of prediction for the seasons. The ancient Egyptians, for instance, based their calendar on the rising of Sirius, the brightest star in the sky. At the time of the year when the star rose just before dawn, this signaled the annual flooding of the Nile. Although astrologers continue to use the stars to make predictions, modern day astronomers make no such connections.

As for Aries, the Sun passes through it between April 15 and May 13. Six months later it is exactly opposite the Sun and can be seen rising in the night sky as the Sun sets.

Taurus

Rising beyond Orion in the winter sky is Taurus, one of the most widely known constellations in the ancient world. Legends of this region of the heavens abound. The Greeks believed that Zeus disguised himself as the bull, Taurus, so as to carry away the Princess Europa of Tyre to the island of Crete. Other stories suggest that Hercules tamed the Bull as one of his twelve labors.

The outline of the Bull's head is strikingly marked by the V-shape of the star cluster, the Hyades. His eye, Aldebaran, which glows red orange, is the brightest star in the constellation and is some forty times larger than the Sun and over one hundred times as bright. The Arabic name for this star literally means "follower," as it chases the clusters of the Hyades and the nearby Pleiades, or Seven Sisters. In mythology, the Hyades and the Pleiades were sisters, the daughters of Atlas, but Aethra was the mother of the Hyades and Pleione was the mother of the Pleiades. Legend has it that Orion pursued the Pleiades, who were nymphs of Artemis. She turned them into stars and placed them in the sky for safety, while she in turn became Orion's lover.

It is said that the Hyades took care of Zeus' child, Dionysus, for which they were rewarded with their positioning in the heavens.

Both the Hyades and the Pleiades are groups of stars all born at approximately the same time and position. The Hyades, though, were formed some 600 million years ago compared with the youthful Pleiades, born a mere 50 million years ago.

Close to the Bull's horn lies the Crab Nebula, site of one of the most explosive events in stellar history. In 1054, Chinese astronomers noticed an unexpected flaring of a star that was so bright that it was visible in daylight for three weeks. What they had observed was a supernova explosion. A supergiant star, near the end of its life, had torn itself apart in a cataclysmic explosion, hurtling material from its outer layers into the void of interstellar space. Today, almost a thousand years after the event, all that remains of the once enormous supergiant is its core material, which has been crushed to form a tiny neutron star, only nine miles in diameter, but with the mass of the Sun. This highly dense star is now spinning rapidly and emitting radiation as it nears its final moments.

Gemini

As Orion and Taurus sink into the western horizon, the constellation of Gemini, the Twins, is still high in the winter and spring skies. It is one of the twelve constellations of the zodiac, with the Sun appearing to pass through it from the end of June to the end of July.

Astrologers arrange the twelve signs of the zodiac into four groups: Earth, Fire, Air and Water, which stem from the ancient Greeks, who believed that these were the only four elements of nature. Gemini is termed an air sign.

The two brightest stars in this region are Castor and Pollux, the heads of the two interlocked brothers. According to legend, these were twins born of Queen Leda of Sparta. Castor's father was Leda's husband King Tyndareus, while Zeus, who visited Leda in the form of a swan, was the father of Pollux. The brothers were both on the crew of the Argonaut, which searched for the Golden Fleece, and traditionally the twins have been regarded as the protectors of sailors.

Although Castor and Pollux appear close to each other in the sky and are of similar brightness, they are really sep-

arated by several million miles. Pollux, an orange giant, is some thirty-six light years from the Earth while Castor, which appears to be blue-white in color, is forty-five light years away. In fact, Castor is an amazingly complex object. A small telescope can identify Castor as two stars that travel around each other over a long period of some five centuries. With some difficulty, astronomers can also detect a third reddish star, which is a dwarf star. These stars are known as Castor A, B, and C. However, the story does not end there. Further examination reveals that each of these components is in fact comprised of two further stars that dance around each other in a matter of days.

Each year in winter, a shower of shooting stars, or meteors, appears to emanate from the constellation. The shower is known as the Geminoids, and it reaches its peak around December 13-14, when as many as fifty per hour can be seen streaking across the night sky. Meteors are not actually stars at all, but minute specks of dust burning up in the Earth's atmosphere.

Cancer

While the great hero Hercules was fighting the many-headed Hydra, guardian of Amymone's well, as his second labor, he was attacked by the Crab, Cancer. Hercules crushed the Crab under his foot, but it was later placed in the heavens.

Cancer is the fourth sign and the fifth constellation of the zodiac, and the Sun appears to pass through this region of the sky between the end of July and the end of August. Astrologers linked the shy, retiring nature of the creature to those born under its sign.

Although relatively small and with few bright stars, Cancer is still interesting, mainly for a beautiful, if faint, star cluster lying deep in its heart. This group has perhaps fifty stars at a distance of some 520 light years. Through binoculars, the stars appear to be buzzing around a honey pot, the center of the cluster, giving rise to its popular name, the Beehive Cluster. In ancient times, though, it was known as the Manger and the two stars that stand on guard by it were called the Northern and the Southern Asses.

The Greek poet Aratus, who wrote the earliest remaining complete description of the sky, mentions the Manger as a forecaster of weather. If the cluster cannot be seen while other nearby stars are in view, then rain is predicted, as even a small amount of moisture in the air is enough to obscure this very faint group.

In the time of the ancient Greeks, the Sun was in the constellation of Cancer when it reached its most northerly point in the sky. This day is known as the summer solstice, as the Sun appears to stand still for a short time before apparently turning south again. At this point the Sun appears directly overhead at midday at the latitude of 23.5° north of the Equator, a line of latitude that became known as the Tropic of Cancer. Since then, however, the situation has changed. The Earth wobbles a little as it turns on its axis, rather like a spinning top. Although it is rather a slow wobble, over thousands of years its effects become noticeable. One particular result is that the annual positions of the Sun have changed and the summer solstice now occurs when the Sun is in the constellation of Gemini. The Tropic of Cancer, though, has retained its original name.

Leo

As spring approaches, Leo, the Lion, can be seen high in the southeastern skies. It is the fifth sign and the sixth constellation of the zodiac, with the Sun traveling through it from August 7 to September 14. It is sometimes thought to be the lion from Nemea that Hercules slew as one of his twelve labors.

Leo is one of the few constellations that bears a fair resemblance to the shape it is supposed to represent. The lion's head is outlined by a group of six stars called the Sickle, with the last star, the brightest in the group, being Regulus, or the Little King. Four quite bright stars mark his body, and his tail is given by Denebula, which, literally, means the lion's tail.

Regulus, the twenty-first brightest star in the skies, is a brilliant blue-white star. Its color indicates that it is very hot, probably with a surface temperature of 20,000°F, compared with the Sun's surface temperature of around 11,000°F. Lying some ninety light years from the Earth, Regulus is five times as large as the Sun and pours out over two hundred times as much light.

Within the Sickle group of Leo, the second brightest star, Algieba, or the mane of the lion, appears as a fairly ordinary star, but when looked at through even a moderate-sized telescope it splits into two beautiful golden stars. These are a pair of orange giants which travel slowly around each other, taking some six hundred years to complete a full turn.

Just east of Leo is a shimmering little group of stars known as Berenice's Hair, or *Coma Berenices*. Berenice was a Queen of Egypt and wife of Ptolemy Euergetes. Once, when he was in battle, she promised the gods that if he returned home safely, she would cut off all her long hair and offer it to Venus as a sacrifice. The King duly returned and Berenice was true to her promise. However, her tresses later disappeared from the altar in the Temple of Venus. A court astronomer, hoping to avoid any trouble, pointed to a little patch glistening in the skies close to Leo and said that the gods were so pleased with the offering that they had taken the hair and placed it in the heavens for all eternity.

Virgo

In the month of May, the second largest constellation, Virgo, can be seen high in the southern skies. A zodiacal constellation, the Sun passes through this area from mid-September to early November.

Several legends have grown up around the Virgin. Some see her as the goddess of justice, Astraeia, holding the scales of justice, represented by the nearby constellation of Libra, in her hand. More frequently, though, she is seen as a maiden, Demeter, goddess of the harvest, and she is depicted holding an ear of wheat. The latter is marked by the fifteenth brightest star in the sky, Spica, a brilliant blue-white star. Spica is twice as hot as the brightest star in the sky, Sirius, but because it lies some 300 light years away it appears dimmer. Sirius, in contrast, is only nine light years away.

Looking much deeper into Virgo we find a great cluster of over 3,000 galaxies lying at the vast distance of forty-five million light years away. Each of these galaxies contains hundreds of thousands of millions of stars. The sheer numbers of stars is truly awesome. In the same way that

stars in our own galaxy, the Milky Way, are often clustered together, so also when we look outside it we see that some of the other great star cities are grouped together. Astronomers believe that our own Milky Way may be part of an enormous super cluster centered on the Virgo cluster.

Staggering though the Virgo cluster is, there is another highlight to be found in this region. The brightest quasar yet discovered lies in this direction, although it is not connected to the cluster. Quasars must rank as one of the most mysterious of all astronomical phenomena. Looking like ordinary stars, they emit huge amounts of energy. Close examination reveals that they lie outside our own galaxy and in fact at vast distances. Since their brightness is so enormous they cannot therefore be regular stars. In galactic terms they are quite small, just about the size of our solar system, and are racing away from us at speeds of over 90 percent of the speed of light. This suggests that they are at the very furthest reaches of our universe, and may hold clues to its very structure. Some astronomers speculate that it is black holes that power these objects and that quasars are, perhaps, the deep cores of galaxies in an early stage of their lives.

Hercules

Hercules is one of the most ancient of the heroic figures of the night sky, and the legends surrounding him link the constellations of Leo, Cancer, Cygnus, Aquila, Hydra and Draco, the Dragon. This giant is often represented as kneeling in the sky with his foot crushing the head of the Dragon, though sometimes he is seen brandishing a club and at other times shooting an arrow from his bow. The Greek version of the name is Herakles, but he probably dates back even further to the Babylonian legends, in which the constellation was portrayed as the Kneeler, who had killed a dragon. As the Greeks extended their influence, it is apparent that they also took on the legends and myths of their surrounding countries and simply replaced the names with those of their own gods and heroes.

Hercules was set twelve labors, or tasks, by Eurystheus. One of these was to bring back the skin of the ferocious lion of Nemea, represented in the sky by Leo. Hercules used all his weapons to no avail, and finally had to kill the beast with his bare hands. A second labor was to kill the

monster, Hydra. The pictures that show Hercules shooting his bow connect his arrow with the small constellation of Saggita, while the birds he is shooting at are Cygnus, the Swan, and Aquila, the Eagle. These are thought to relate to another of his labors, which was to kill the birds of the swampy Stymphalian marsh.

For all the heroic tales, this constellation lacks really bright stars, although it is an area rich in double stars. The brightest star in the group is known as Ras Algethi, which is a red giant. This is a phase in the latter stages of a star's life, when it swells to an enormous size while becoming rather cool at its surface. Ras Algethi is thought to be one of the largest stars known, although in common with other stars of its age, it varies in size by pulsing in and out occasionally. At some time in the future our own star, the Sun, will move into this phase and will then expand out to the orbit of Mars, swallowing the Earth as it does so. Fortunately, this is not likely to happen for another five thousand million years.

Libra

Libra, the Scales, is a small, faint constellation of the zodiac, through which the Sun passes during the month of November, and is the only inanimate zodiacal member. It is visible during the summer months in the southern regions of the skies.

At present, Libra is represented by a set of scales. However, this is a constellation which has gone through several changes. The ancient Greeks referred to the stars in this group as Chalae, the Claws, referring to the claws of the nearby Scorpius. In fact the two brightest stars in Libra are still known as Zubenelgenubi, the Southern Claw, and Zubeneschamali, the Northern Claw. However, during the time of Julius Caesar in the first century BC, the Romans designated it as a separate constellation. The Scales are now associated with the goddess of justice, Astraeia, who is sometimes represented by the nearby figure of Virgo.

Cygnus

Cygnus, the Swan, is not only one of the most beautiful constellations, with its bright summer stars shining brightly overhead, it is also one of the most engaging, with a wealth of fascinating astronomical puzzles.

The ancient Greeks saw Cygnus as a swan flying down the Milky Way. Legend has it that Zeus visited Leda, the wife of Tyndareus, King of Sparta, in the form of a swan. He seduced her and she gave birth to Pollux, one of the Gemini twins.

The constellation is marked by a cross-like shape, and is often called the Northern Cross. Its tail is given by the extraordinarily bright Deneb, its neck by the star Albireo, and its wings by two slightly dimmer stars. Deneb is a massive blue-white supergiant, pouring out as much light as 80,000 suns. It is only because it lies far from Earth, at 1,700 light years away, that it does not dominate the sky even more.

Within Cygnus lies the bizarre star known as P Cygni. This star has been carefully watched for centuries, as it has been seen to brighten dramatically and then fade away again. It is thought to be highly unstable, and is spewing out millions of tons of its outer layers in occasional bursts. It is a very likely candidate for a supernova explosion, which would rip it apart, flinging its remnants over vast distances in space.

Cygnus is also the home of a suspected black hole. A black hole is an immensely dense object, probably the final stage of a star that has totally collapsed on itself. The force of gravity associated with it is so strong that anything coming close to it will be sucked into it, with no possibility of escape. Not even light can make its way out of its depths. Since it is not possible to "see" a black hole its presence can only be detected by inference. Within Cygnus is a powerful source of X-rays. Astronomers believe that a black hole, has formed a pair, or binary, with another fairly ordinary star. However, as this pair orbit around each other, the black hole is stripping off the outer layers of its companion. As the material is dragged towards the black hole, the material is rotating so quickly that it gives off X-rays. This is rather like a last cry for help as, once the material is within the hole, nothing more can be detected.

Scorpius

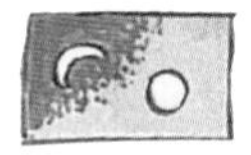

Scorpius, also known as Scorpion, is one of the truly ancient constellations of the night sky, inspiring tales in many different cultures. In Greek mythology it was the scorpion who killed the mighty hunter Orion, with a fatal sting, and to this day Orion is always seen in the opposite part of the sky from the creature, forever fleeing from it.

The constellation looks remarkably like a scorpion, with a curve of stars marking its stinging tail, and its heart designated by a dull red star, Antares. The name of this star means "the rival of Mars," Mars being a very distinctive red-colored planet. However, Mars gets its hue from the large amounts of rust on its surface reflecting sunlight, while Antares is actually emitting red light from its cool outer layers. The constellation has been reformed since ancient Greek times. The scorpion's claws have been taken away from the creature and now form the scales in the nearby constellation of Libra.

Scorpius is the most southerly of all the zodiacal constellations. However, because of the effects of the slight

wobble in the Earth's motion (precession), the Sun does not reach its winter solstice position in this group but in the neighboring constellation of Sagittarius. The winter solstice is the most southerly point in the Sun's apparent annual motion. In fact, the Sun passes only briefly through Scorpius for a week or so at the end of November.

The zodiac is further complicated by the presence of another large constellation, Ophiuchus, which lies between Sagittarius and Scorpius. As the Sun spends several weeks in this group in December, it appears that there are in fact thirteen signs of the Zodiac rather than the traditional twelve.

Scorpius is also the home of the strongest known X-ray source in the cosmos. We are used to looking up at the skies and seeing visible light from the stars. However, radiation of all types, including ultraviolet, infrared, and radio waves, can be picked up from all parts of the universe. Since it seems unlikely that we will ever travel to the stars, all our understanding of the heavens must come from the little scraps of radiation that we pick up on Earth. One of the greatest challenges to astronomers is to detect these clues and then pull them all together to try and make sense out of this perplexing but beautiful universe in which we live.

Sagittarius

Sagittarius, the Archer, is one of the Zodiacal constellations. His upper part is depicted as a man with a bow and arrow raised, ready to shoot at the nearby Scorpion, while his lower half is a horse. It is the constellation where the Sun reaches its most southerly point from the Equator on the day of the winter solstice, December 22.

Sagittarius lies in the path of the Milky Way, in one of its richest stretches, with many clusters and nebulae.The Milky Way is the name given to the magnificent radiant stream that stretches right across the night sky, looking like a river of milk. It is also the name that we give to our own galaxy. This is a gigantic collection of stars grouped together in a huge spiral shape. As we live in the outer region of the spiral, when we look towards the center we see so many stars that their light blends together to look like a milky path.

Pegasus

As the autumn nights of the northern hemisphere draw in, the skies reveal a rich set of constellations, including Pegasus, Andromeda, Cassiopeia and Perseus, which are all linked together through the epic tales of Greek mythology. Pegasus, the winged horse, was the steed of Perseus. This amazing animal was born of the blood of the terrible Medusa, whom Perseus killed.

The most distinctive part of the constellation is the Square of Pegasus, outlined by four moderately bright stars. Very recently, in 1995, astronomers detected for the first time what they believe may be a planet outside our solar system. It appears to be traveling around the third brightest star in the Square. If this is the case, it opens up the possibilities of other life forms in the universe; if not in this constellation, then perhaps in others.

The region is most famous for an enormous globular cluster. This cluster has hundreds of thousands of very ancient stars, all tightly packed together in a spherical form.

Andromeda

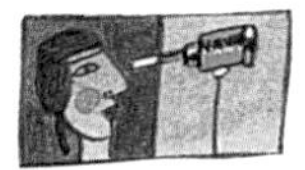

Andromeda was the daughter of the Ethiopian Queen, Cassiopeia and the King, Cepheus. She ran into trouble because of the conceit of her mother: Cassiopeia, boasting that she was even more beautiful than the sea nymphs, so enraged them that they persuaded Poseidon, the ruler of the waters, to send the terrible sea monster, Cetus, to prey on the country's coastline. So great was the destruction that after some years Cepheus, in despair, sought the advice of an oracle. He was told that the country could only be saved if his daughter, Andromeda, was chained to a rock and offered as a sacrifice to the monster. Andromeda was within the jaws of the beast when the timely intervention of Perseus saved her. He was flying past on his winged horse, Pegasus, when he saw her distress. He succeeded in cutting her loose, then he killed Cetus and took Andromeda as his bride.

Andromeda lies beside the constellation of Pegasus. In fact, the blue-white star, Alpheratz, which is one of the four stars comprising the Square of Pegasus, more properly belongs to Andromeda as it marks her head.

In contrast to the thrilling legends of the region, the constellation does not appear particularly distinctive at first glance. Astronomers must use powerful instrumentation to look into its depths and savor its delights, the most wonderful feature of which is the presence of our nearest galactic neighbor. Technically known as M31 but more usually referred to as the Andromeda galaxy, the light from this vast collection of stars takes just over two million years to reach Earth. It is the most distant object in the night sky that our eyes can distinguish.

By looking at this great spiral form, astronomers are able to get a feel for the shape of our own galaxy, something that is normally difficult to do, given that we live within it. The Andromeda galaxy contains 300,000 million stars and appears to be a little larger than our own Milky Way. It also has two smaller satellite galaxies that travel with it. It was the discovery of M31 that first led astronomers to realize the vastness of the universe, that there were, in fact, countless other galaxies of stars and that the Milky Way was but one, rather ordinary, galaxy.

The Great Bear

Ursa Major, the Great Bear, is one of the most distinctive of all the constellations, and the third largest. Many ancient peoples in different lands and cultures have visualized the group as a bear, although some Europeans saw it as a chariot, and the Arabs thought of it as a coffin. Nearby is the smaller and dimmer constellation of the Little Bear, Ursa Minor, which has a similar shape to the Major group.

Greek mythology tells a tale of Zeus seducing the nymph Callisto and then transforming her into the Great Bear. Close by, her son Arcus is portrayed as the Little Bear. Another legend suggests that they represent the nymphs Adrasteia and Io, who took care of Zeus when he was a child and who were then placed in the sky as she-bears. Incidently, Io and Callisto are also the names given to two of the satellite moons that travel around the planet Jupiter.

Ursa Major is a circumpolar group, which means that from high Northern latitudes it can be seen throughout the year as it moves around the North Pole Star, Polaris. Its seven brightest stars are also known as the Plough or the

Big Dipper, which make up the rump of the Bear. Within the saucepan, or bowl part of the constellation, five of the stars are related to each other. They would have been born around the same time and at one point would have been much closer together. It appears they are now slowly drifting apart.

One of the most interesting stars in this group is Mizar, at the bend of the Dipper's handle. Close by is Alcor, and these two are known as the Horse and Rider. Until recently, a well-known test of good eyesight was to distinguish these two stars. Mizar, however, is a binary star. Its dim companion is sluggishly orbiting Mizar, taking about 10,000 years to complete a full turn.

The Dipper is one of the great signposts of the night sky. Visible throughout the year, the last two stars in the bowl section are known as Merak and Dubhe, or the Pointers, since a line drawn through them will point directly towards Polaris in Ursa Minor. A line drawn in the opposite direction points to the constellation of Leo, the Lion, while following the arc of the handle of the Dipper will lead to the red-orange star Arcturus in the constellation of Boötes, the Herdsman.

Cassiopeia

From the northern latitudes of Britain and North America, the constellation of Cassiopeia is always visible in the night sky and lies on the opposite side of the Pole Star from the constellation of the Little Bear. It has a very easily distinguishable M or W shape, although the traditional picture of the group is of the proud and haughty Queen Cassiopeia sitting regally in a chair. Cassiopeia's boastfulness drove the sea nymphs, the Nereids, to take revenge. They convinced Poseidon, King of the Seas, to send the sea monster Cetus to terrorize the coastline, and humiliated Cassiopeia by causing her to continually circle the Pole Star.

This region is particularly rich in stellar clusters, although most lie so deep in space that astronomers require powerful telescopes to detect them. Some, however, can be picked up using binoculars, and reveal hundreds of sparkling stars. One of the greatest mysteries in Cassiopeia is its immensely strong radio source. Scientists believe that it may be a fragment from a supernova explosion in the seventeenth century, although there is no record of such an

event at that time. Only by piecing together scraps of information gathered over the years will astronomers eventually be able to resolve the puzzle.

The third brightest star in Cassiopeia is situated at the center of the W, and has kept astronomers intrigued for years as, periodically, it seems to brighten dramatically and then fade away. The star is known to be a very hot blue star, probably twice as hot as our Sun and, amazingly, 5,000 times as bright. We think that it is spinning rapidly, and that this leads to shells of material being literally blown off the surface of the star, causing the fluctuations in brightness. This material is shed at speeds of hundreds of miles per second and will eventually make its way into the interstellar clouds. At a much later date it will be recycled to form a new set of stars. Astronomers reckon that our Sun, and indeed the Earth itself, were born from star dust that had been part of two or maybe three other stars. These stars lived and died long before our solar system came into being, and presumably remnants of our system will likewise form the basis of future stars in millions of years to come.

Perseus

Perseus, the great hero and the son of Zeus, stands astride the northern regions of the sky. Perseus's grandfather had been warned that his daughter's son would cause his death, and so he shut both of them in a chest and threw it into the sea. They were miraculously saved by a fisherman, and Perseus went on to feature in many of the ancient legends, invariably carrying out noble deeds.

It was Perseus who took on the frightful Medusa. She was one of the three gorgons, with serpents for hair. So fearful was Medusa that anyone who even looked at her was quite literally petrified, or turned to stone. Perseus cunningly avoided her gaze by looking only at her reflection in his shield. He cut off her head and from her blood sprang the winged horse, Pegasus. Perseus continued on his travels and subsequently rescued the unfortunate Princess Andromeda, who was chained to a rock and just about to be devoured by the sea monster, Cetus. Perseus kept the Medusa's severed head in a bag, sometimes thwarting his enemies by producing it and so turning them to stone. The

constellation is occasionally depicted with Perseus holding the severed head, and the star Algol, which means "the ghoul," is said to represent this fearsome object.

Algol has its own peculiar properties, appearing to wink balefully at its observers. It varies in brightness every few days, and analysis of its starlight has revealed that it is a binary star, that is, two stars revolving so closely around each other that they cannot be seen separately. As the larger one passes in front of the other, part of the light is obscured or eclipsed, and so Algol appears to be dimmed until this component passes. The eclipse lasts for eighteen minutes every three days. The two stars of Algol could hardly be more different in nature. The brighter of the two, Algol A, a beautiful white star, is a hundred times more brilliant than the Sun, while Algol B, the larger of the two, is a faint orange star.

Like Cassiopeia, this region sparkles with many clusters of stars. However, the most fascinating of these groups is marked by Perseus' hand as he brandishes his sword. This is known as the Double Cluster, and has two families of stars. Although these clusters seem to be side by side one of them is actually 300 light years further away. Yet again, our eyes play tricks on us.

Capricorn

Capricorn, the Sea Goat, is represented as a goat with the tail of a fish. It has been associated with the god Pan, who is sometimes depicted as having a goat's head. Legend has it that Pan dived into a river to avoid the monster Typhon and, to hasten his escape, he turned the lower half of his body into a fish.

It is the tenth sign and the eleventh constellation of the zodiac, with the Sun in occupation from January 18 to February 14. Over two thousand years ago, the Sun was in Capricorn when it reached its furthest point south of the equator on the winter solstice, December 22. The Sun appears directly overhead at noon at the latitude 23.5° south on this day, and this line became known as the Tropic of Capricorn. It still retains the name, even though the Sun is now in the constellation of Sagittarius on the day of the winter solstice.

Aquarius

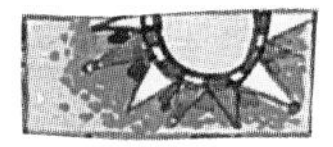

Aquarius, the water bearer, is the eleventh sign and the twelfth constellation of the Zodiac. It has been chronicled since the most ancient times. The Babylonians imagined an old man pouring water from a jar into the mouth of a fish. This fish is represented by the nearby constellation of Piscis Austrinus, or the Southern Fish, not to be confused with the twin fish of Pisces.

In Greek mythology. the figure was a young shepherd boy by the name of Ganymede, who was appointed by Zeus as the wine waiter to the gods and sent to Mount Olympus to serve them.

The water bearer's jar is clearly marked above four bright central stars, which are the most easily distinguished in this fairly small constellation.

The Sun is in Aquarius from late February to early March. Hence this group will rise above the horizon in the night skies some six months later.

Pisces

Pisces, the Fishes, is the last of the zodiacal constellations. It appears to represent two fish, sometimes shown as being tied together either by their tails or by a ribbon.

At present, the vernal equinox occurs when the Sun is in the constellation of Pisces. This is the point when the Sun is directly overhead at midday at the equator, and marks the annual point when the Sun moves from the southern celestial hemisphere into the north. This day is usually March 23. Because of the effects of precession, this point will move into Aquarius in about 600 years' time, and this will mark the beginning of the Age of Aquarius, a period noted by astrologers. In the time of the ancient Greeks, some 2,500 years ago, this point was in Aries and indeed over the next 25,000 years, as the Earth continues to wobble on its axis, the point will move through all of the zodiacal constellations.

Glossary

binary star	Two stars orbiting around each other.
black hole	Extremely dense object which gives rise to a very strong gravity field.
color	The color of a star depends on its surface temperature. Blue-white stars are hottest, red stars are coolest.
constellation	Imaginary grouping of stars, often associated with a particular pattern or picture.
double star	Two stars which appear close together. Some may be widely separated in space, others are true binaries and orbit each other.
dwarf star	Smaller than average star, sometimes the last stage in the life of a star.
galaxy	Gigantic collection of stars held together by gravity.
globular cluster	Spherical cluster of old stars, found at outskirts of galaxy, containing hundreds of thousands of stars.
light year	Distance light travels in a year (5,880,000,000,000 miles).
meteor	Shooting star, a tiny piece of cosmic dust which burns up as it enters the earth's atmosphere.
Milky Way	Name given to the galaxy in which we live.
moon	Small body orbiting a planet.
nebula	Misty-looking patch in the sky, usually made of huge clouds of dust and gas.
neutron star	Tiny, rapidly-spinning star, the end product of a supernova explosion.
open cluster	Group of between 10 and a few hundred stars. As they drift apart, the cluster "opens."

planet — One of nine bodies orbiting the Sun.

precession — An effect causing a wobble in the spin of the earth which gives rise to a change in the apparent positions of the stars over thousands of years.

quasar — Immensely powerful object traveling at high speed at the edge of the universe. Looks like a star, but produces more light than many galaxies combined.

solar system — Collection of celestial bodies whose motion depends on the Sun.

star — Very large ball of burning gas. Nuclear fusion at the center produces its light.

stellar cluster — Group of stars within a galaxy, all born around the same time. Two types exist, open and globular.

sun — The star closest to earth, of average size and brightness. It is about 100 times bigger than the earth.

supergiant — Extremely large star, hundreds of times larger than the Sun.

supernova — Star which has exploded, tearing itself apart.

zodiac — Collection of constellations through which the Sun appears to pass each year. In fact it is not the Sun moving, but the Earth as it orbits the Sun.

List of Constellations